Harry Thomas Cory

The Thompson-Houston System of Electric Lighting

Harry Thomas Cory

The Thompson-Houston System of Electric Lighting

ISBN/EAN: 9783337249649

Printed in Europe, USA, Canada, Australia, Japan

Cover: Foto ©berggeist007 / pixelio.de

More available books at **www.hansebooks.com**

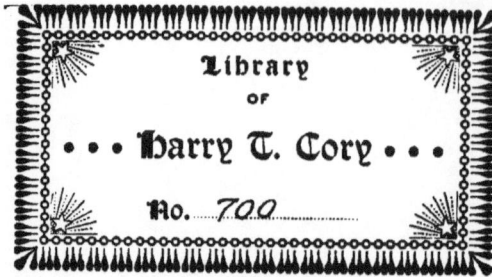

The Thompson - Houston

System of Electric

Lighting.

Thesis submitted

for the degree of

Bachelor of Science

in Mechanical

Engineering,

to the Faculty of Purdue University

June 1887

"In its power to assume always that form of energy which happens to be the most useful lies the great importance of electricity: This importance has been brought to the notice of the public by means of the many recent exhibitions. Public interest has been roused and there is everywhere a desire for information and a guide through this far reaching field for discovery and invention. And, although there are many works treating on electricity and electric light, people specially want a short and concise though thorough description of the various schemes by which electric light is produced. In this thesis the object is to give a brief treatise on one of the many schemes of producing light by electric currents viz - The Thomson-Houston system.

In pursuing the subject of electrici-

1., the first thing noticed is the analogy
and difference between the dynamo
and its older and more powerful rival
the steam-engine. The resemblances are,
First as in the developement of the steam
engine, but few of the improvements and
inventions in electrical machines were
made by mathematical leaders. Watt
ran across the idea of the seperate
condenser while repairing the New-
comen model and applied the
expansion of steam to the steam
engine by a mechanical accident
rather than by his own ingenuity.
and so we find the first designers
of the dynamo were mechanics rather than philosophers. Secondly the
tendency to disregard old methods
and instruments because of new
discoveries and inventions has, as in
the steam engine. hindered the ad-

vancement in electrical science.
As an example it has become custom-
ary to regard frictional and statical
electric machines, for practical
purposes, as obsolete, but recent discoveries seem to hint that they may
yet be utilized. Lately Prof. Dodge has
shown that dust and vapor whirl-
ing in the air may be settled by a
discharge of electricity consisting of
a continuous series of electric sparks
This has been utilized to clear the at-
mosphere in lead smelting works
from the fumes of volatized lead
and with its application comes the
invention of Wimhurst which produces with a minimum of mechanical labor a continuous series of electric sparks and works admirably.
       The differences between the en-
gine's and dynamo's development are:

First the marvelously rapid developement of the dynamo as compared with
that of the steam engine. Since 1867
when the term "dynamo electric machinery" even to scientific men had
but little signification, the dy
namo has been brought to a very high
degree of perfection. Secondly, the
development of the dynamo has
reached a much higher degree of
perfection than that of the steam
engine. Among the best steam en-
gines twenty per cent efficiency is
considered as very good while a good
dynamo gives out in the form of
electricity, ninety per cent of the
mechanical energy put in it. But
the class of people who improved and
made the steam engine what it is
were as well educated in one sense
as were the men who brought out

the dynamo. While it is true that in Watt's time the knowledge concerning steam was very meagre, yet the practical men who made the dynamo, did it by themselves as nearly all the teachers of electricity knew nothing except what may be called electrical tricks. As has been said "The teachers and writers of textbooks, practically did not know that there was anything in common between the electricity from a rubbed glass machine and voltaic electricity, or to be brief, that there was a science of electricity as distinguished from mere natural history." In fact as late as 1870 there were really no textbooks.

— · — · —

* The foreg; statement is quoted from Dr Urbitzkany's work "Electricity in the Service of Man."

on electricity. Even now electrical knowledge is so meager as to warrant the same writer's expression, "We can not imagine a mechanical engineer mistaking a few inches for a few miles or a grocer confounding an ounce of sugar with a carload, but this gives too truthful an idea of the vagueness that still exists"

In the distant future, electricity will be used for electric lighting only as subordinate to other uses to which it may be applied such as heating houses, taking place of stoves for cooking, being used as a substitute for the steam engine. In fact the motor is rapidly becoming of as much practical use as the electric light. The principle of the motor is just this; a certain amount of mech-

anical energy say thirty four horsepower per minute into the form of
electric currents, which by the way
gives enough current to run 45,
2000 candle power lamps, send the
current and distance through suitable conductors and attach them
to similar dynamo or dynamos' but
in such a manner that the current
in the second set of dynamo's flows
in the reverse direction to that
of the first; when, the armature of
the second dynamo or dynamos will
revolve and at the pulley or pulleys
of the dynamos, aside from friction,
will be given out 95% the thirty-
four horsepower, the loss being
due to the resistance of the conductors. Now in practice a motor
is placed on the arc light circuit
the same as a lamp, for energy

less than twelve horsepower. It does not affect the lights and is a clean, neat way of obtaining energy.

But however true the foregoing may be, the greatest present use of electricity is to start and maintain light. There are several so-called systems, embracing dynamos, lamps, regulators, etc. from which I select the Thomson-Houston as the one for the purpose of describing for several reasons, first, it is at least as good as the average system of which there is a mushroom growth; second, valuble information was, kindly offered by the parent Company; third, a good plant is near to which free acess was given, and fourth, we have at the Mechanical Hall of this University, a dynamo, loaned by the parent company,

which affords information without
any inconvenience. As each part
of the system comes up to be described a little of its history will be giv
en. As the first part of a system
necessary to be produced is the current generator we will first descib.

The <u>Thomson-Houston</u> <u>Dynamo</u>.

In considering the current generator the first thing to be decided
upon is the definition of the term
dynamo. The following is thought
to be a correct definition,—A dynamo or dynamo electric machine is
a machine which is used to con-
vert energy in the form of mechanical motion into energy of electric
currents, or vica-versa. Those used
to generate currents of electricity
are called dynamos, those used to
generate mechanical motion are

known as motors.

In attempting to make clear the theory of the dynamo, we will recall some simple experiments. In Fig. 1, send a current around B from right to left. Now A being free to move vertically either up or down, connect its binding posts to a galvanometer (that is, an instrument used to tell the direction of a current and also used to test the relative strength of two or more currents) and move A up suddenly when a current will be generated in A whose direction will be the same as that of the current in B. Now this current is not created energy, because in lifting* the coil A, work is expended§ against the attraction

---

* Gravity does not enter, as a current is generated in lowering A.

between the coils, as between two currents flowing in the same direction. there is an attraction. If we pursue this experiment in its various forms we will find the following statement known as Lentz law is true, viz: "If the relative positions of two conductors A and B be changed of which B is traversed by a current, a current is induced in A in such a direction that by its electro dynamic action on the current in B it would have imparted to the conductor a motion of the contrary kind to that by which the inducing action was produced"

The theory of this laws is that around every wire carrying a current there is a magnetic whirl (Fig. 0). Now if the conducting wire be passed through a hole in a horizontal plate of glass and iron

filings be sifted upon the latter
they will arrange themselves,
as shown in Fig. 2, along lines, ...
dial in this case, known as lines
of force, which arranging is due
to the magnetic attraction of the
current in the wire upon the
iron filings. Now in B. Fig. 1, every
portion of the wire has just such
a whirl and just such lines of
force, or magnetic field, and when
A is moved each part of the wire of
A cuts one or more lines of force of
the many magnetic fields making up the magnetic field of the
entire coil B.   Now when the wire
of coil A cuts magnetic field of B a
current is generated in A according
to the following statement known
as Faraday's Law; "When a conductor
in a field of force moves in any way

so as to cut the lines of force then is an electromotive force produced in the conductor in such a direction that supposing a figure swimming in the conductor to turn to look along the positive direction of the lines of force (in Fig. 1, toward axis of B) and the conductor be moved to his right, he will be swimming with the current so induced. Hence in Fig. 1, the current generated in A will be from left to right.

Practically Faradays principle means just this: by moving a wire across a space when there are magnetic lines, the motion of the wire as it cuts the magnetic lines sets up around the cutting wire a magnetic whirl or in other words sets up a current in that wire.

The forgoing laws are the "prin

principles of the dynamo," yet after
their deduction, the progress of the
evolution of the dynamo was slow
and attended by many dificulties.
Between 1860 and 1870 however, a work
ing knowledge of these laws became
the property of thousands of mechanics, and by comparing the number of inventions before and after
that date (1860) the present generous
growth of systems, dynamos and
lamps, prove that inventions were
almost in proportion to the number of people who had any electrical knowledge. In 1866 Wilde produced a toy magneto-electric machine
for giving shocks, in which he used
excited electromagnets. In the same
year Varley and others produced a
machine which excited its own
field magnets the type of all ma-

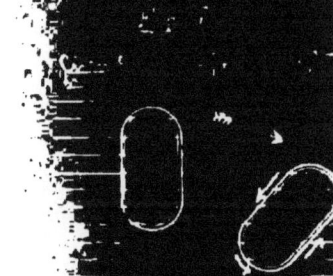

chines used in practice. With this
principle of Varleys and Pacinnotti's
ring, Gramme produced in 1871 his
since famous continuous current
generator, one of which the second
dynamo electric machine ever brought
to this country can now be seen at
the engine house at Purdue University.
In 1877 Silas Brush brought out his
famous dynamo and it may be interesting to know that he designed
and had one made without experi-
menting in the least. In the following year a patent was issued to
Messrs Elihu Thomson and Edwin J.
Houston, Professors of electricity in Phil
adelphia on the present though much
improved, Thomson Houston Dynamo

To go back to Lentz and Faraday's
laws and carefully consider them we
can but assent to S. P. Thompson's "Fifteen

propositions on the dynamo" which
are :- 1. A part of the energy of an electric current exists in the form of a magnetic whirl surrounding the wire

2. Currents may be generated in a
wire by setting up these whirls.

3. We can set up these whirls by
increasing or decreasing the relative distance between magnets and wires

4. To set up and maintain these
whirls consumes power.

5. To induce currents in a conductor there must be motion between
them so as to alter the number of
lines of force (Fig. 4 to 7)

6. Increase in the number of
lines of force in the circuit produces
a current of the opposite sense to
decrease (Fig. 7)

7. Approach induces electromotive force in the opposite direction

to that induced by retreat

8. The stronger the magnetic field the stronger the current.

9. The more rapid the motion the stronger the current.

10. The greater the length of conductor which cuts lines of force the stronger the current.

. 11. The shorter the conductor not so employed the stronger the current.

12. Approach being a finite process the approaching and receding must give alternating directions to the current

13. By the use of a commutator all the currents can be turned in the same direction

14. In a steady circuit it makes no difference what kind of magnets are used to procure the requisite magnetic field whether permanent

or electromagnets

15 Hence the current of the generator may be used to excite the magnetism of field magnets.

Now the Thomson-Houston dynamo comes under that class of dynamos in which there is a rotation of coils in a uniform field of force, such rotation (Fig. 6.) being affected round an axis in the plane of the coil. Of course this dynamo is made like all others of its class to have _first_ as powerful field-magnets as possible _second_ the armature or rotating coil has as great a length of wire in it as possible the wire being thick to offer little resistance and _third_ built to stand high rotation speed

The simple theoretical dynamo is shown at Fig. 8, consisting of a single rectangular loop of wire.

rotating in the magnetic field formed by large magnets, and in order
to take the current so generated from
the loop so as to give a continuous
current, we use a two part commutator (Fig. 9) consisting of a metal
tube split in two and mounted on
wood, each half connected to one end
of the loop. The current is taken off by
brushes which lead to the main circuit. But manifestly this dynamo
would give no appreciable current
becase it has a very small length
of wire on the armature, so a great
number of loops were used which
at present constitute the so-called
drum armature.

   We may rotate the loops of wire
in Fig. 8, on one of its sides as an axis or even push it farther from the
center of revolution than that. To do

this, wrap the wire around a ring and connect both ends to a two part commutator (Fig. 10). If instead of the ring in Fig. 10, being solid it be a number of coils of wire and if instead of there being one coil around the ring there be thirty we will have Pacinnottis ring before spoken of. If we used four to ten coils or "bobbins" of large size which is shown diagramatically at Fig. 11, we would have the Brush dynamo.

So with exceptions we may say that there are practically two types of dynamos as regards armatures, the ring type as Brush, Pacinnottis &c., and the drum armature (page 20)

The Thomson-Houston dynamo is like the rest of that dynamo, unique. To quote S. P. Thompson; The Thomson-Houston spherical armature is unique among armatures. its

cupshaped field magnets are unique among field magnets, its three part commutator is unique among commutators."

An armature of a dynamo is the rotating coil or coils which generates currents of electricity by moving in a magnetic field of force. It is the most important part of a dynamo as it is literally the current generator. So we first consider the Thomson-Houston

Armature

It is spheroidal in shape as is noted for the fact of its very seldom burning out, i.e. the electricity heating the wires of the armature to such an extent as to destroy the insulation or fuse the wires, either rendering the armature useless. It is made by keying two dish-shaped iron disks S S Fig 12, to the shaft x and putting ribs dd about

ten in number in the twenty-five
light machine, and over the whole
putting varnished paper. Then at stated intervals, pegs JJJ are driven into
suitable holes in the disks and ribs
to help in winding wire on the shell.
Next three insulated wires of equal length
are joined together at h Fig. 13 and the three
wires are then wound over the shell
in the following peculiar manner:
one half of No. 1 is wound so as to form
a zone of a sphere of which the shaft is in the
same plane as the center circumference
of the zone. The armature is then turned
on the shaft as an axis 120° and one half
of No. 2 is wound in the same manner
as the first half of No. 1. The armature
is moved 120° more and all of No. 3 is
wound. The armature is then turned
back, 120° on the shaft as an axis and
the remainder of No. 2 is wound. Lastly

the armature is turned back 120° more
and the rest of No.1 is wound. They are
bound by wires gg Fig.13 to hold them
when rotating. The object of this rather com-
plicated winding is to get the three coils
equi distant from the shaft in order that
each coil will generate practically the
same current. Now as will be seen the
overlapping wires will form a nearly spherical armature. The armature is moun-
ted on the shaft $\underline{x}$ as an axis which ex-
tends far enough out from its bearings
to put a pulley on the end $\underline{H}$ and a commutator on the other end to the three
parts of which are fastened the three
wires marked one, two three, Fig.13

It has been urged that the repairs
of this armature will be larger than
on any other armature. If there should
be a 'burnout' it would necessitate the
taking apart of the dynamo and send

ing the armature to the factory to be
rewound. But it never burns out except
through positive carelessness and it
will be found that the repairs on this
armature is less than on the armatures of its several powerful rivals
taken separately even though they be
of simpler construction.

When the Thomson-Houston armature is rotated between the cup shaped
field magnets alternate currents are
generated in each coil in turn and
now the next point to be considered is
the

Commutator

which unites the alternate currents
so formed into one continuous current.
The commutator as before stated is
fastened on the shaft at the end one, two
three Fig. 13. It consists of three copper
plates in the form of a cylinder each eg.

ment A'A'A covering 115° of the dotted circle
Fig.14. They are screwed to rod CCC and DDD
which are insulated by wood and gutta-
percha plates EE from the iron mounting
E'' which is in turn screwed to shaft by
set screws shown. The wires one, two, three
have respectively red, white and blue
insulation and are put in bindingposts
DDD marked one, two, three at the factory and if not so placed may work badly
The current enters D goes to DD' which then
have direct contact with A.

Now in a three part commutator the
spark occurring as the segments pass under the brushes would very quickly destroy
the surface and interfere with the currents in the coil. This difficulty is overcome
by blowing out the spark by an air blast
given at just the right place and time. The
manner in which the blast is delivered is
as follows: the segments of the commutator

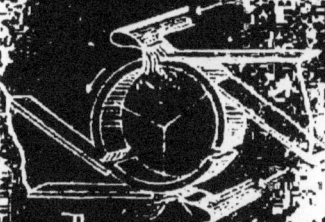

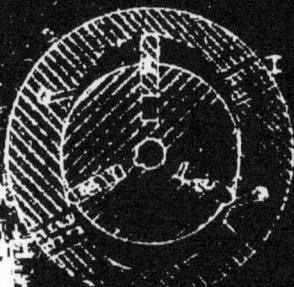

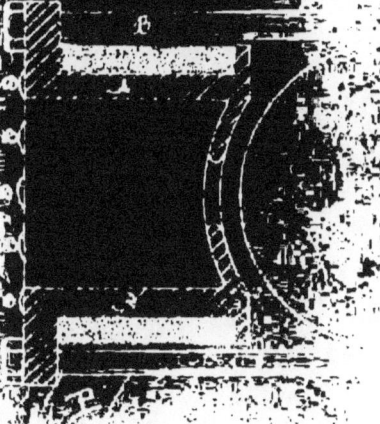

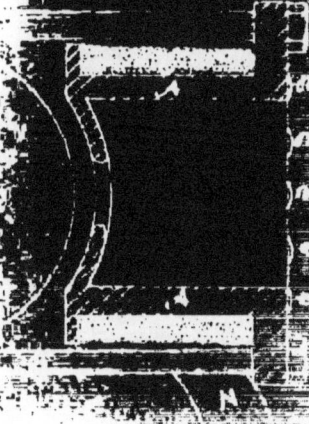

are separated by gaps of about 5° and in front of each of the leading brushes there projects a nozzle, Fig. 15, which discharges an air blast alternately three times in each revolution. The blast itself is supplied by an ingenious piece of mechanism known as the

Thomson Air Blast

It consists of an elliptical box II whose sides have perforations II when air can enter while inside of this rotates a steel disk keyed to the armature shaft and having radial slots in which slide three wings RRR of ebonite which as they fly around drives air into the holes JJ leading to the nozzles Fig. 15. The result is that, since the spark is done away with, oil can be supplied to the commutator in limited quantities but still amply sufficient to reduce the wear on the

commutator to such an extent that the
life of a segment is greatly increased.
The air blast is fastened to the dynamo
frame just behind the commutator and
can be seen in Fig. 23.

The Field-Magnets,

as may be seen from Fig. 17, consists
of two flanged iron tubes AA whose end
consists of a convex segment of a sphere
accurately turned to receive the armature.
Coils of wire CC which are in the outside
circuit and through which the entire
current flows are wound upon the
tubes. After the armature is placed
between them the two tubes are bolted
together by heavy wrought iron bars
BB and the whole carried on the frame
work PN shown also at PN Fig. 23. Now,
a little magnetism only remains in the
wrought iron bars and iron ~~frame~~ frame
works when the armature first re-

valves, but the current even though slightly
going through the coils makes an electromagnet out of each tube and heavily
magnetizes the wrought iron bars and
in two or three seconds after the armature first rotates it is entirely surroun-
ded by a heavy magnetic field. One of
the good points of these field magnets
is that but very little magnetism is lost
as compared with most other dynamos
and since it takes power to maintain
a heavy magnetic field this dynamo is in
this respect very economical.

The Thomson Regulating Gear

Later on we will show that pushing the
brushes together or pulling them apart alter the strength of the current, but for
the present just accept the fact and
we will show how the brushes are varied.
It is accomplished by the mechanism
shown in Fig. 18. The brushes are fixed to

the levers YY and Y_1Y_1 united by the lever l. The automatic movement is obtained by the electromagnet R while a dashpot J prevents too sudden motion. Suppose the brushes to be in the position shown when the current would get too strong owing to lights being cut out. The electromagnet R getting stronger would raise A and reduce the current taken off until current came to normal. If, instead, some lamps were thrown in the current would become weak and the electromagnet R would become weak, drop A which would increase current and this will continue till current reaches normal.

The foregoing regulating gear is used on small dynamos and old style large ones. On the large new style dynamos a more delicate regulating gear is used, the current which operates it being shown at Fig. 19. Normally the electromagnet R is

short circuited by the wire T and only acts when this circuit is broken. At some point in the main circuit is a wall controller or controller magnet shown in Fig. 19 at ST. consisting of two electro magnets their yoke supported by a spring and the yoke operating the contact lever S. If the current becomes too strong the controller magnet circuit is broken and all the current of the main circuit goes through the electro-magnet R which by its sudden increase of strength quickly raises A and thus alters the brushes. This only exists for a moment until the yoke of the controller magnets fall because of their decrease of magnetic strength, when current again flows through wire T because when yoke drops contact is made This decreases the strength of electro-magnet R thus dropping A and increasing current. Hence S will again raise and break contact and R again raise A This is contin

ually repeated.

The Brushes,

of which there are four in use on all machines, are made of a broad strip of springy copper having six slits two thirds the distance up, and thus touching at several points. They are held by clamps shown at Fig. 23 which also shows the brushes. The brushes are held to the commutator by their own springiness and the variation of position due to strength of current. The brushes are set by a gauge sent with each dynamo which shows length from the end of brush to the holder. The holders are set at the correct angle by a gauge of brass of the shape of a right angled triangle the short side having a wide flange curved to fit the commutator for which it is sent, while the second side are re-

gards length must fit, to the holder when
swung to it on the commutator as
an axis.

After describing the details of
the dynamo, we will at once pro-
ceed to find how the

~~Thomson~~ Houston Dynamo
Operates

In the diagram Fig. 19 the rotation
is as in practice against the hands
of the watch when seen from the commutator end of shaft. The three coils
of the armature are represented by
three lines A, B, C, united at their in-
ner extremities each being joined
to a segment of the commutator. There
are two positive brushes P and F and
two negative ones P' and F'. The current
delivered to P and F goes round one of the
field magnet coils, then to the outer
circuit consisting of regulating gear,

lamps, motors, etc, through the other field magnet coil to brushes P' and F.' Now from Fig. 20, we observe that supposing the loop to be rotating against the hands of the watch in a magnetic field the diagram represents by arrows the direction of the electromotive forces induced in those loops. The action is a maximum along the line of the resultant magnetic field $m\,m'$ and the minimum along the line $n\,n'$ which is at right angles to $m\,m'$. The reason that $m\,n'$ is not horizontal is that the induced poles of the armature is in advance of the poles of the field magnet and is constantly tending to be drawn back. Applying Fig. 20 to Fig. 19, we see that there will be an outward current in B, an inward one in C, A generating no current for that moment.

207111

Now the following pair of brushes FF' are shifted backward three times as far as PP' is shifted forwards. When the current is the greatest possible the brushes P and F and P' and F' are 60° apart thus leaving P and F' and P' and F' just 120° apart and since the segments of the commutator are each 120° in length* there will always be two coils in parralel with one another and in series with the third. Taking one sixth of a revolution and continuing all the way round we find the following tabulated statement showing brushes in contact with coils, to be true viz: -.

---

* Each segment is really only 115° in length but the brushes are set at a distance from the holder far enough to just reach over the five degree gap by the gauge above described.

From external circuit $\left\langle \dfrac{P-C}{E-A} \right\rangle B \left\langle \dfrac{P'}{F'} \right\rangle$ to external circuit

    "   "   " $\left\langle \dfrac{P}{F} \right\rangle A \left\langle \dfrac{P'-B}{E'-C} \right\rangle$ "   "   "

    "   "   " $\left\langle \dfrac{P-A}{F-B} \right\rangle C \left\langle \dfrac{P'}{F'} \right\rangle$ "   "   "

    "   "   " $\left\langle \dfrac{P}{F} \right\rangle B \left\langle \dfrac{P'-C}{E'-A} \right\rangle$ "   "   "

    "   "   " $\left\langle \dfrac{P-B}{E-C} \right\rangle A \left\langle \dfrac{P'}{F'} \right\rangle$ "   "   "

    "   "   " $\left\langle \dfrac{P}{F} \right\rangle C \left\langle \dfrac{P'-A}{E'-B} \right\rangle$ "   "   "

Now suppose the current to become to strong owing to any cause, the following brushes are made to recede. This can but shorten the time that the brushes are in contact with the commutator when the coil is passing through that position in which it is generating the maximum amount of current and also hasten the time when it goes into parralel with a comparatively idle coil. If the current is to weak then the brushes are made to close up thus reducing the time that the most active coil is in parralel with one less active and also makes the

brushes be longer in contact with the segment when the coil is generating its maximum amount of current. The motion of advance and retreat of the brushes is accomplished by the Thomson Regulating ~~Gear~~ before described. On Fig. 23 can be seen all the dynamo's details except the Controller Magnet.

As regards the ~~Thomson~~-~~Houston~~ ~~Dynamo~~ it will be found to produce the steadiest and most uniform current of any dynamo now in use. Its regulating gear is the simplest and most natural one ever used. In its ability to reduce the current simultaneously to one tenth of its former quantity inside of one or two minutes ~~without~~ ~~injury~~ to ~~itself~~ and ~~lamps~~ it stands alone, in practice.

—11—

Fig. 23.

Your engraving
representing
"Dynamo Electric
Machine
with Thomsons
Spherical Ar-
mature"

}

‑ Taken from one of
your catalogues.
and pasted on a
sheet of this paper—

———— " —— " ————

In a system the most important thing next to the dynamo is the lamps. The first experimenter who produced an electric glow was Otto von Guericke. But neither the glow nor electric spark have been used to produce electric light for practical purposes, this was left to the voltaic arc on the one hand and the incandescent lamp on the other. Davy in 1800 mentions experiments in which electric light was obtained by electric sparks between two carbon points. He showed the arc* light for the first time in 1810 at the

————————

*An arc light is a light produced by the use of the voltaic arc, which is made by the sparks passing between two poles of a powerful battery which are brought together and then separated a little

Royal Institute, which with Foucalt's hand regulator (1844) Deleuil lit the Place de la Concorde, Paris. Thomas Wright in London (1845) devised the first apparatus which automatically adjusted the carbons. W. C. Staite used the electric current for the regulation of the carbons in 1848. In 1855 Serrin constructed a lamp which would have been used on a large scale had it not been for the cost of generating electricity. In 1876 Paul Jablochkoff invented his electric candles and in 1881 there were 4000 in use, but as their use increased their defects were found out. Regulated lamps were again brought into use and with them experimenters again endeavored to solve the problem of dividing the electric light. In 1877 Tschikoliff solved the problem in a very simple

manner. He reasoned that, if the current br divided and part go through the carbons and make the arc and the rest go through an electromagnet and regulate the arc and the the current unite and when another light is wanted the current br again divided and reunited, the current may br divided any number of times and the scheme work nicely. When put in practice it worked very nicely and is used on most lamps at present Suppose there br a lamp placed in the circuit. The current divides and the larger half goes through the carbons, as here there is no resistance as the carbons touch, while the remainder, going through a spiral of high resistance, is small. When the carbons burn away a little the arc is formed and the resistance

increasing brings the regulating gear
into operation. Now the strength
of the current is the same after it
has gone through the lamp as before
because the current is going to get
through either one way or the other,
hence any number of lamps may go
on in series, depending only upon the ten-
sion of the current.

Incandescent lamps were produced as early as 1859 but not till
1879 when Swan, Edison, Sawyer and
others were they ever in a practical
form. The first glow lamp Edison
constructed had platinum wire
to be heated. He however examined the
properties of organic substances and
finally fixed on bamboo fibre. The bam
boo is divided into fibres one millime-
ter in diameter and twelve milli-
meters long. These fibres are pressed

in U-shaped moulds and baked in o
vens where they are allowed to become
carbonized. The carbonized filament is
attached to platinum wires which
are fused in a glass vessel from
which the air has been exhausted. We
will speak more fully of the incandescent lamp when describing the Thom
son Houston System's incandescent
lamp.

The Thomson Arc lamps was u-
sed by the Thomson Houston System
since its begining till about two years
ago when they stopped manufacturing
them, only furnishing broken parts.
The arc lamps at present used is –

The Thomson Rice Arc Lamp.

They are manufactured in two styles
the single lamp used for stores, buil-
ings etc, and the double lamp used
for street service, all night work, etc.

The light is produced by the voltaic arc between two carbons, the negative pole or lower carbon burning away about half as fast as the positive pole or upper carbon. The outside view of the single lamp is seen in Fig. 21 and of the double lamp in Fig. 22

The regulation of the double lamp is diagramatically shown in Fig. 24, which is a plan of the lamp with cover removed, showing only a plan of cylindrical part of the lamp. The wires marked $\underline{a}$ $\underline{b}$ $\underline{c}$ $\underline{d}$ run along the top in order to be out of the way. In Fig. 24 the current comes in at the binding post and is at $\underline{A}$ divided into three currents $\underline{A}$, $\underline{B}$, and $\underline{c}$ The current $\underline{a}$ goes to the yoke $\underline{I}$ of the electromagnets $\underline{h}$ and $\underline{i}$ and where the yoke is not held down by magnets $\underline{h}$ and $\underline{i}$, it goes out wire $\underline{a}$ to binding post $\underline{B}$. This only continues

a moment until the current b which goes through the carbons and at the start has almost no resistance offered it, attracts the yoke I thus breaking contact of circuit a until the current ceases or till both carbons burn away, when in the latter case the resistance of b becoming very high as compared to j and k but little current goes through h and i and I is raised by a weak spring not shown, thus making contact of circuit a, and since current a has little resistance as compared to b or c most of the current goes through it, thus practically making a cut-out. The current b goes round the electro-magnets h and i, then to the "bed" through screw J, the "bed" being a cast iron bottom of the cylinder E Fig. 22. From the bed it goes down carbon holder C (or H) through carbons and arc to frame bed A Fig. 22. From

H. T. Cory

there it comes up a wire by the side of frame _C_ Fig. 22 and joins other currents at _B_. The third current _c_ goes through electromagnets _j_ and _k_ and joins other currents at _B_.

This is when switch _F_ Fig. 22 and _M_ Fig. 24, is turned on. Now since the dynamo will regulate all differences in current the lamps can be turned on or off at will by any one. This is accomplished at the lamp by turning off the switch. When the switch is turned off, the current goes through _a_ to screw _K_ which is then touched by metal _L_ (in contact with binding post _B_ and worked by _M_)

It will be perceived that any disorder in a lamp cannot affect other lamps in the circuit and will right itself or if not the lamp can immediately be switched out of circuit.

Now as to the regulating gear. The two carbon holders are held up, H, by clutch operated by springs (not shown) till end N of lever ON is permanently held down, and C, by the raising and falling of yoke D. There is only one arc burning at a time in a double lamp and the socalled positive carbon C burns first. When the lamps are trimmed the switch is first turned off the carbons put in and the switch turned on. This will draw the upper carbons up about a quarter of an inch.

When the current is turned on the circuit aa is almost instantly broken and most of the current goes through c as the distance between carbons being a quarter of an inch the arc has a very large resistance. The electro magnets of ank attract D which lets loose C, which falls to lower carbon, and the

resistance being almost nothing, most of the current goes through b. This weakens j and k which lets D up while D takes c up with it thus establishing the arc. The current all goes down C till the enlarged end of C strikes lever ON thus letting H drop and also putting it in electrical contact with "bed," which it was not in before. After a short time the carbons burn away, the arc becomes longer and establishes itself and the resistance becoming greater in passing from carbon to carbon and a correspondingly less current flows through b and a greater one through c. This makes the electromagnets j and k strong enough to draw D to them in spite of spring Q. When D is attracted by j and k C (or H) falls and again the arc lengthens, always being kept about 3-32 inch long. This is frequently and continually repeated, the delicacy depend

ing upon the strength of the spring Q as compared to the electro-magnets strength.

When the carbon in carbon holder C burns to a length of about two inches in attempting to fall to maintain arc's length, an enlarged part at the top of the carbon holder C strikes and holds down lever ON pivoted at Q (and end N held up by a spring P) thus letting loose a clutch by which electrical contact is made between N and "bed" and letting N fall till it touches lower carbon when an arc is established and regulated just as for C.

The Thomson-Rice single lamp has the same gear with the exception of having only carbon holder C, N, lever ON, and spring clutch and spring P being absent. The single lamp will burn eight hours and the double lamp fourteen –

hours continuous running.

These lamps are intended only for a steady current and will not cut out of circuit if the current gets too strong But with the Thomson Houston dynamo the current never gets too strong and because of this there are less power absorbing mechanism and as anythings functions decrease the remaining functions are increasedly better. As the Thomson-Rice lamp has less functions and power consuming machinery, it can but be the most economical, delicately adjusted and steadiest lamp extant. They are made to stand a current of five amperes above the normal current for a short time, as, when forty lights are simultaneously cut out of a forty-five light circuit, the current runs up about four amperes above the normal current for about one half a minute.

Prof. Thomson has gotten out a divided arc lamp which supplies a light of moderate candle power for locations where a 2000 candle power lamp gives more light than can be economically utilized. It is specially suited for factory and mill use where looms or other tall machines are liable to cast disadvantageous shadows. It is said that these lights are supplied cheaper per candle power than the standard lamp and up to date is successful

He has also arranged apparatus by which arc lamps are run in multiple series, series or multiple arc. It is said that divisions, redivisions and reunions are practicable. This is also successful as far as we can find out.

The <u>Sawyer-Man Incandescent Lamps</u>.

As before stated Edison fixed up

on carbonized filament of bamboo. The Sawyer-Man company however applied for a patent on carbonized filament for incandescent lamps on January 19th 1880, and after five years litigation with Thos A. Edison they were granted a patent No. 317,676, on May first, 1885, covering their invention. The Sawyer Man lamp Fig. 25 consists of a carbonized connected to platinum wires fused in a glass tube from which all the air possible has been extracted. The light is produced by the glow of the filament and heat of gases given from filament. The life of a lamp is from 1000 to 1500 hrs and requires a current of 1¼ to 1.3 amperes and give 20 to 25 C. P. When the filament becomes brittle and breaks the tube is unscrewed from the key Fig. 26 and a new one screwed in. They are run on the arc light circuit by the use of an

Fig. 25

consisting of drawing of Sawyer Man lamp cut from catalogue, and trimmed to contour of drawing

Fig. 26

a drawing showing action of key in Sawyer Man lamp, cut to contour

a drawing of the Thomson Rice Individual distributor cut from catalog and pasted in.

Fig. 27

a drawing cut from pamphlet showing, "Method of using Thomson Rice Individual distributor"

Fig. 28

individual distributor Fig. 27 which consists of a brass case containing a magnet in the circuit of the lamps and a resistance coil automatically substituted in case the lamp should break or is turned off by key Fig. 26. The scheme of arranging lamps so as to get the right current is shown at Fig. 28. the number of lamps in a group depending on the current

Prof. Thomson has gotten out a lamp Fig. 29 in two styles one for 6.8 amperes current and one for 10 amperes current. Three lamps of different candle power, due to different potential differences at binding post of lamp, are used on the same current. The method of connecting them is shown in Fig. 3. It will be perceived that the lamps carry the full current yet have a life of 1000 hrs or more. This is a great im

Fig. 29

a drawing
cut from
pamphlet
showing
"Prof. Thom-
son's Series
Incandescent
Lamp—S.I.
Incandescent
Lamp"

Fig. 3o

Drawing showing
"method of using the
Series Incandescent
Lamp manufactured
by the Thomson-Hous.
ton Elec. Co." cut from
your pamphlet
and pasted on
a similar sheet.

vention indeed doing away with a
great loss of power due to high resistance coils. It will be noticed however
that a 125 C.P. incandescent lamp uses
as much energy as a 2000 c.P. arc light,
the 65 C.P. lamp one half as much as
the 32 C.P. lamp one fourth as much

General Remarks

The Thomson Houston system
also furnish lightning arresters, am-
meters, hanging boards, switch boards,
hoods, insulators, lamp arms, etc, but,
though in some respects many of these
miscellaneous articles are ingenious
and novel, yet they are not distinc-
tive of the Thomson Houston or any
other system. Be it said however that
all these articles fill their proper place.
The company also furnish a motor to
go on their circuits but for the double
reason that of the motors not being strict-

A photographer of
La Fayette photoed
the Gas Company's
plant of J & H in this
city one evening at
10 o'clock when several
lights were burning in
room. I had a large one
printed and pasted on
a piece of bristol ~~for~~
board of the same
size as this sheet,
and put in my
original copy.

ly related to electric lighting and of being unable to obtain a description of it, it must remain undescribed as far as this thesis is concerned.

After describing all the parts of the system it may be interesting to know how a plant is arranged. The last plate is a photograph of the LaFayette Gas Company's Plant of the Thomson Houston System taken at ten oclock one night. It shows the engine, dynamos, the wall controller on the left wall, and a view of the lamps which had hoods put before them to prevent the polarization of the negative.

On the accompanying page will be found a table showing experiments with an old style dynamo given Purdue University by the Thomson Houston company which dynamo is now in the engine house of the Mechanical Hall.

Experiments with Three Light T.H. Dynamo No 79

The dynamo was run by a large pulley (about four and one half feet in diameter) on the same shaft as the fly wheel and beside the latter. Two lamps were put in circuit with a Deprez-Carpentier amm-meter and a volt meter of the same make was put in between the brushes. First one lamp (old Thomson style) was switched out of circuit, the dynamo started and when speed was reached the circuit made. The following readings were taken when the engine made 139 + the dynamo 1122 revolutions per minute

Readings At End of	One Lamp		Two Lamps		When 2nd lamp was switched in		When 2nd Lamp was switched out	
	Amp.	Volts	Amp	Volts	Amp.	Volts	Amp.	Volts
1 Second	10	55	10	110	6.7	55	14	110
30 .	10	55	10	110	10	75	10	85
2 Min	10	55	10	110	10	109	10	60
3	10	55	10	110	10	110	10	55

www.ingramcontent.com/pod-product-compliance
Lightning Source LLC
Chambersburg PA
CBHW021819190326
41518CB00007B/668